Saving Our Planet

Saving Earth's Air

by Brienna Rossiter

FOCUS READERS®

PIONEER

www.focusreaders.com

Focus Readers is distributed by North Star Editions:
sales@northstareditions.com | 888-417-0195

Produced for Focus Readers by Red Line Editorial.

Photographs ©: Shutterstock Images, cover, 1, 4, 7, 8, 10, 12, 15, 16, 19 (top), 19 (bottom), 21

Library of Congress Cataloging-in-Publication Data
Names: Rossiter, Brienna, author.
Title: Saving earth's air / by Brienna Rossiter.
Description: Lake Elmo, MN : Focus Readers, [2022] | Series: Saving our planet | Includes index. | Audience: Grades 2-3
Identifiers: LCCN 2021006311 (print) | LCCN 2021006312 (ebook) | ISBN 9781644938218 (hardcover) | ISBN 9781644938676 (paperback) | ISBN 9781644939130 (ebook) | ISBN 9781644939581 (pdf)
Subjects: LCSH: Air--Pollution--Juvenile literature. | Renewable energy sources--Juvenile literature.
Classification: LCC TD883.13 .R67 2022 (print) | LCC TD883.13 (ebook) | DDC 363.739/2--dc23
LC record available at https://lccn.loc.gov/2021006311
LC ebook record available at https://lccn.loc.gov/2021006312

Printed in the United States of America
Mankato, MN
082021

About the Author

Brienna Rossiter is a writer and editor who lives in Minnesota. She loves being outside in the fresh air.

Table of Contents

Chapter 1

Air Pollution

Harmful **chemicals** go into the air. They cause air pollution. The chemicals can be gases. They can also be tiny **particles**.

Some kinds of pollution are easy to see. One kind is **smog**. Other kinds can't be seen. But they are still dangerous. Pollution can make people sick. It can also harm the **environment**.

Did You Know?

Air pollution can cause acid rain. This rain hurts plants and animals.

Chapter 2

Causes

Some air pollution comes from nature. But most of it comes from people. People use lots of energy. To get energy, they burn **fossil fuels**.

Fossil fuels make factories run. They make cars run, too. People also use fossil fuels to heat their homes. When fossil fuels burn, they pollute the air.

Did You Know?

Wildfires send lots of smoke and ash into the air.

ЗОНА

Chapter 3

Problems

Air pollution can be harmful. Many chemicals are unsafe for people to breathe. They can make people sick. They can even cause death.

Plus, some chemicals are **greenhouse gases**. Lots of greenhouse gases are in the air. They are causing **climate change**. This leads to many problems. **Droughts** are more common. So are floods. Storms are stronger, too. Both people and animals face these dangers.

Greenhouse Gases

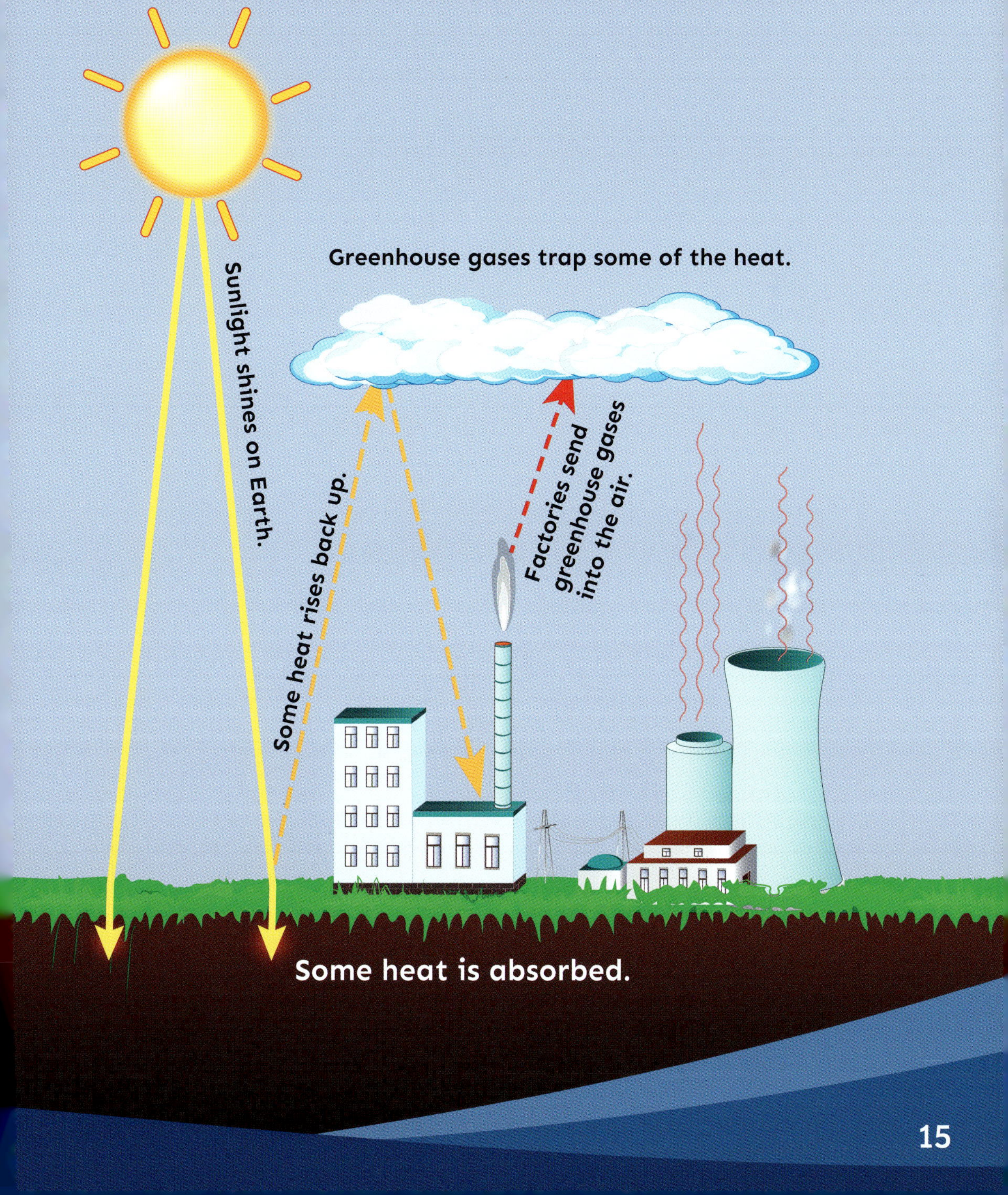

Chapter 4

What We Can Do

People can make changes to protect the air. They can use fewer fossil fuels. They can get energy in other ways.

People can burn less coal. They can get power from wind or sunlight instead. Cities and countries can make these changes. By working together, they can make the air cleaner.

Did You Know?

Pollution from one part of the world can spread to other areas.

SAY YES
to
CLEAN ENERGY

GAS is NOT
Clean or Green

Use Less Fuel

Vehicles are a big source of air pollution. To help Earth's air, people can change how they travel. They can make fewer trips by car. Instead, they can bike more. They can walk more. And they can take buses and trains. These actions use less fuel.

FOCUS ON

Saving Earth's Air

Write your answers on a separate piece of paper.

1. Write a sentence describing the main ideas of Chapter 1.
2. Saving Earth's air involves cities and countries, not just individual people. What can groups do that individuals can't?
3. What is the main source of air pollution?
   A. fossil fuels
   B. wildfires
   C. wind energy
4. How could a drought put people in danger?
   A. People can't see during a drought.
   B. People could have too much water.
   C. People could run out of water.

Answer key on page 24.

Glossary

chemicals
Basic forms of matter. Some chemicals are helpful, and some are harmful.

climate change
A dangerous long-term change in Earth's temperature and weather.

droughts
Long times of little or no rain.

environment
The natural surroundings of living things.

fossil fuels
Energy sources that come from plants and animals that died long ago. Fossil fuels include coal, oil, and natural gas.

greenhouse gases
Gases that trap heat in the air that surrounds Earth.

particles
Very tiny pieces.

smog
A dark, thick cloud of dirty air that hangs near the ground.

To Learn More

BOOKS

Orr, Tamra B. *Air*. North Mankato, MN: Capstone Press, 2021.

Rivera, Andrea. *Air*. Minneapolis: Abdo Publishing, 2018.

NOTE TO EDUCATORS

Visit **www.focusreaders.com** to find lesson plans, activities, links, and other resources related to this title.

Index

Answer Key: 1. Answers will vary; **2.** Answers will vary; **3.** A; **4.** C